CLASSIFIED:

PROJECT:

FILE UNDER:

ULTRA TOP SECRET ROBOT DINOSAUR BLUEPRINTS

BY: DAVID CUNLIFFE

NATIONAL INSTITUTE FOR

ROBOT DINOSAUR RESEARCH

CLASSIFIED:

PROJECT:

FILE UNDER: # ULTRA TOP SECRET ROBOT DINOSAUR BLUEPRINTS

BY: DAVID CUNLIFFE

NATIONAL INSTITUTE FOR

ROBOT DINOSAUR RESEARCH

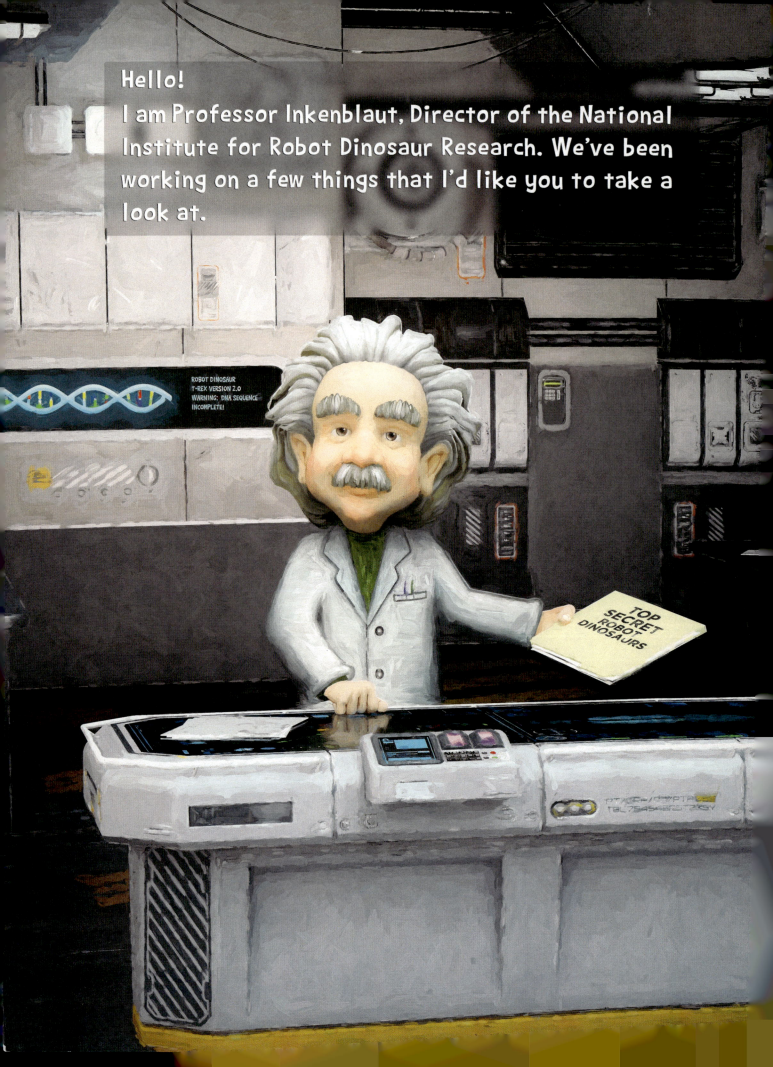

In your hands you hold all of our research on Robot Dinosaurs, including the blueprints to build the prototypes.

Remember, these are Ultra Top Secret, so only share them with those you trust completely!

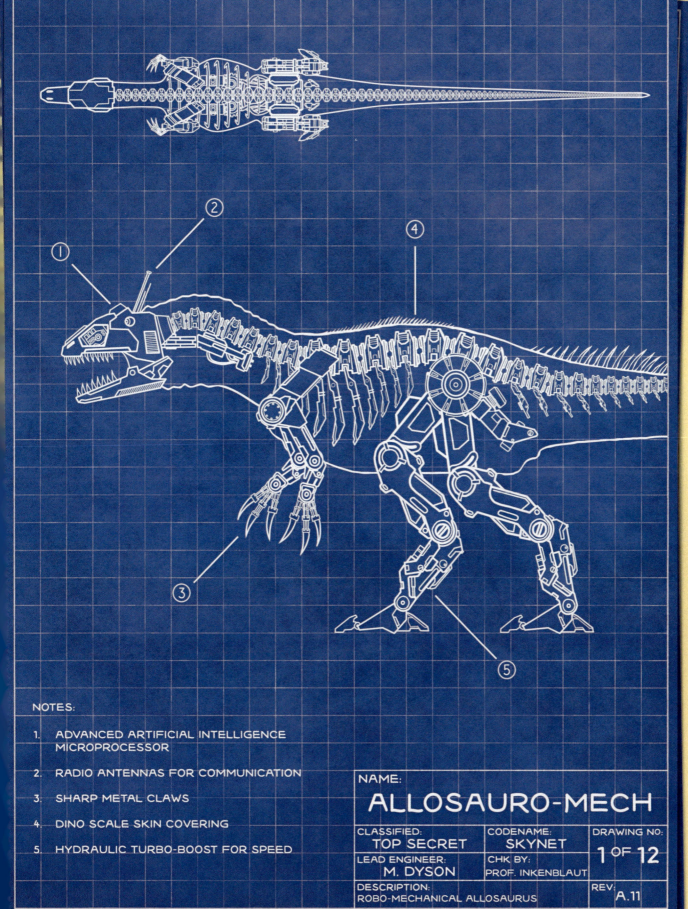

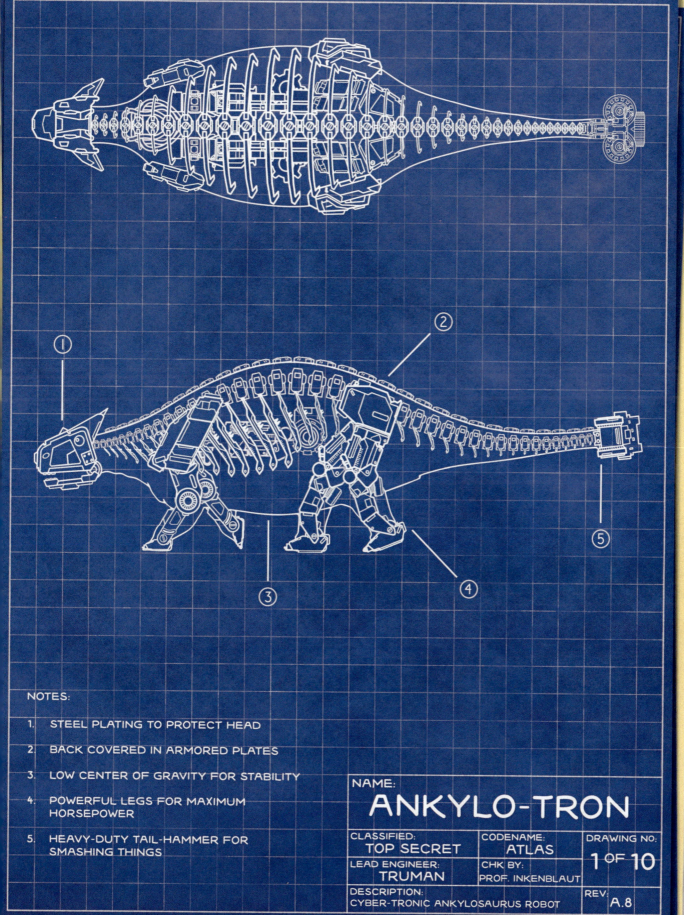

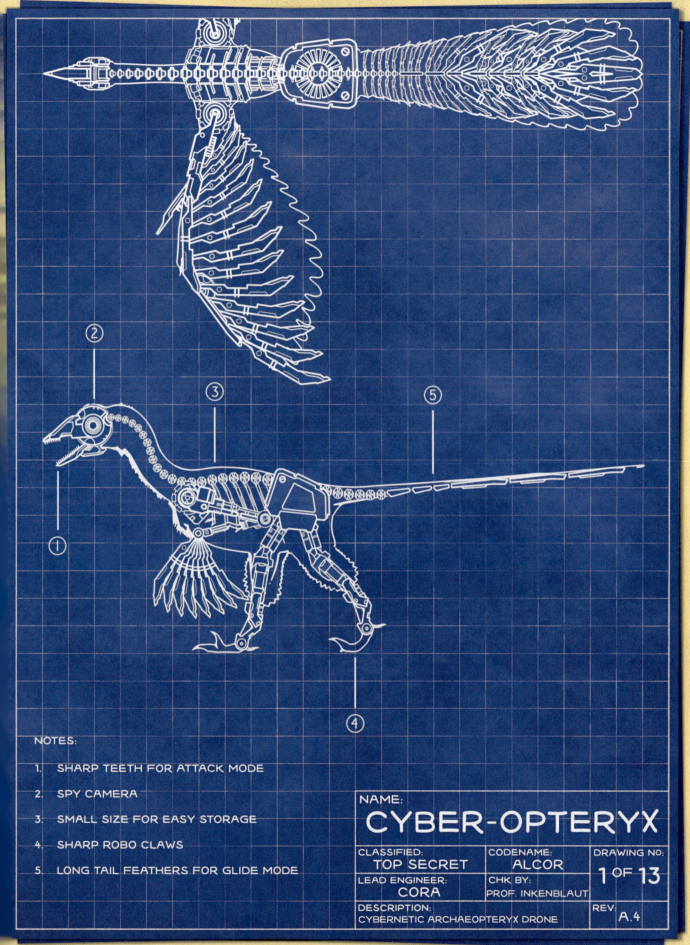

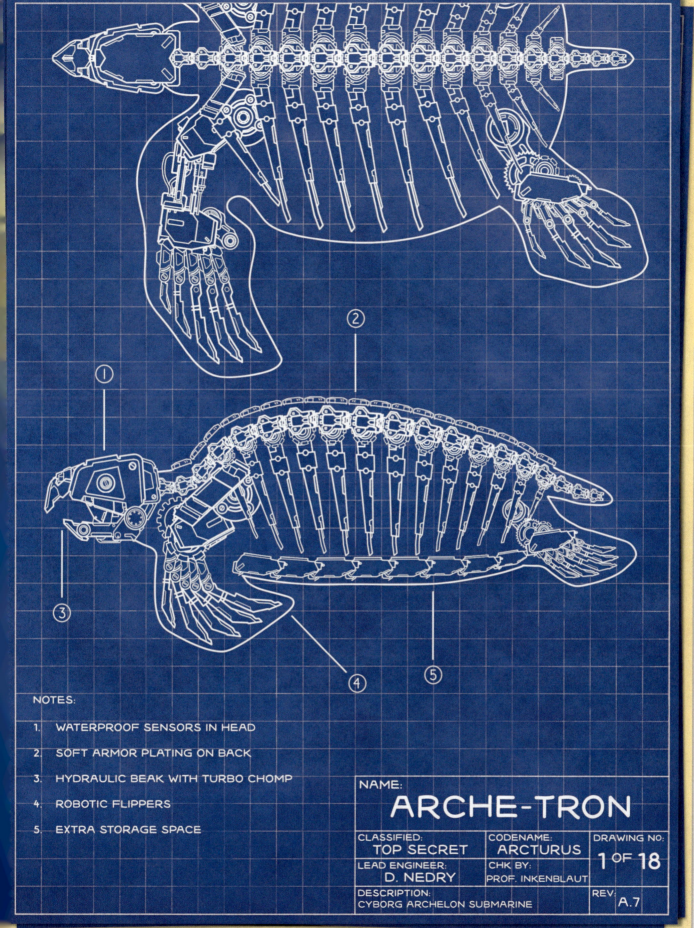

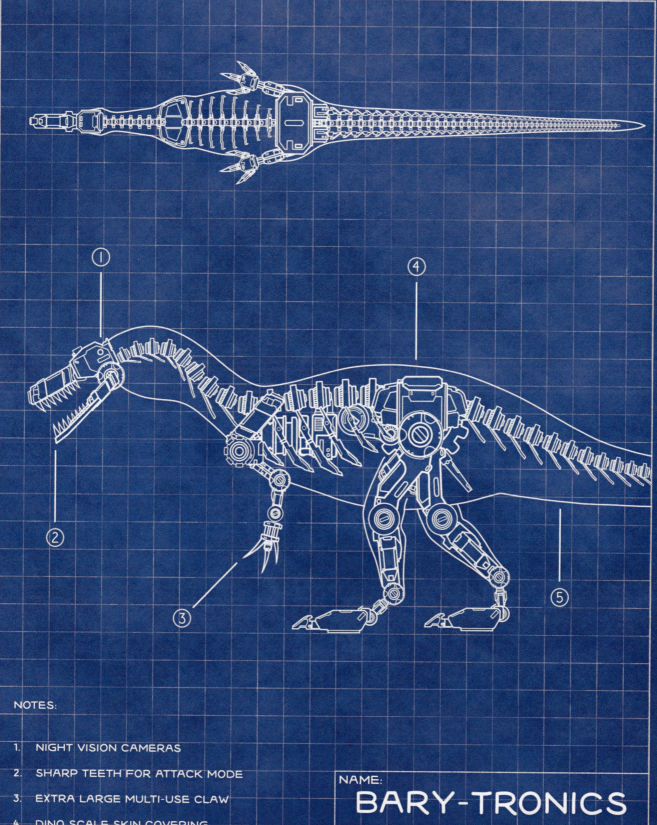

NOTES:

1. NIGHT VISION CAMERAS
2. SHARP TEETH FOR ATTACK MODE
3. EXTRA LARGE MULTI-USE CLAW
4. DINO SCALE SKIN COVERING
5. POWERFUL ROBO-TAIL FOR KNOCKING THINGS OVER

NAME: **BARY-TRONICS**		
CLASSIFIED: TOP SECRET	CODENAME: BRACHIUM	DRAWING NO: **1 OF 16**
LEAD ENGINEER: THEOPHILUS	CHK BY: PROF. INKENBLAUT	
DESCRIPTION: AUTONOMOUS BARYONYX ROBOT		REV: B.5

BARYONYX

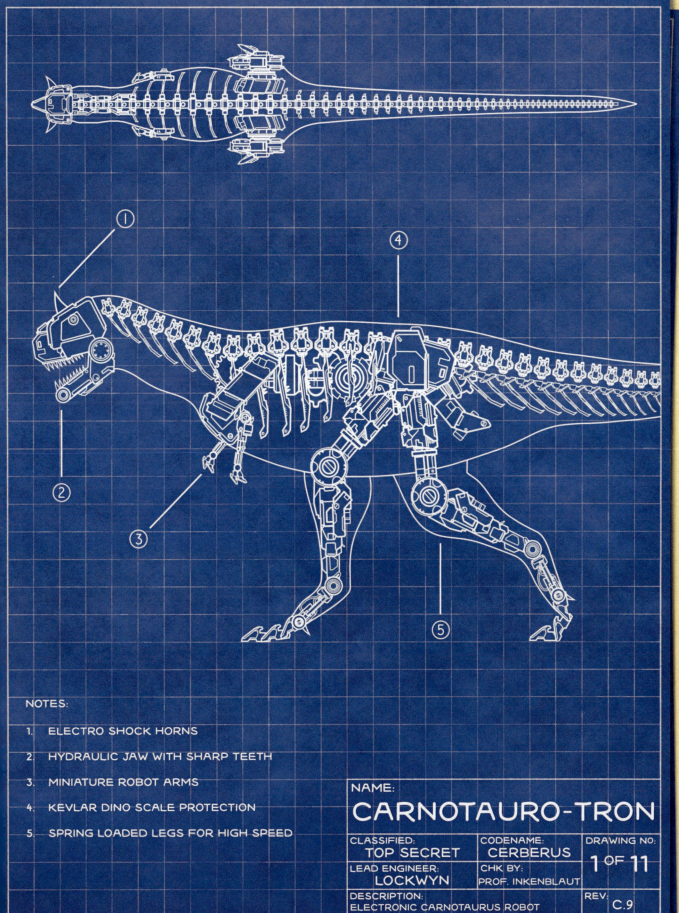

DILOPHOSAURUS
(DIE-LOF-OH-SOR-US)

EARLY JURASSIC: 193 MILLION YEARS AGO

- TWO CRESTS ON HEAD
- LONG TAIL FOR BALANCE
- LONG, SHARP TEETH
- LIKELY HUNTED IN PACKS

TYPE: DINOSAUR
HEIGHT: 6 ft
LENGTH: 20 ft
WEIGHT: 1,000 lbs
DIET: MEAT
HABITAT: RIVERBANKS

20 ft

6 ft

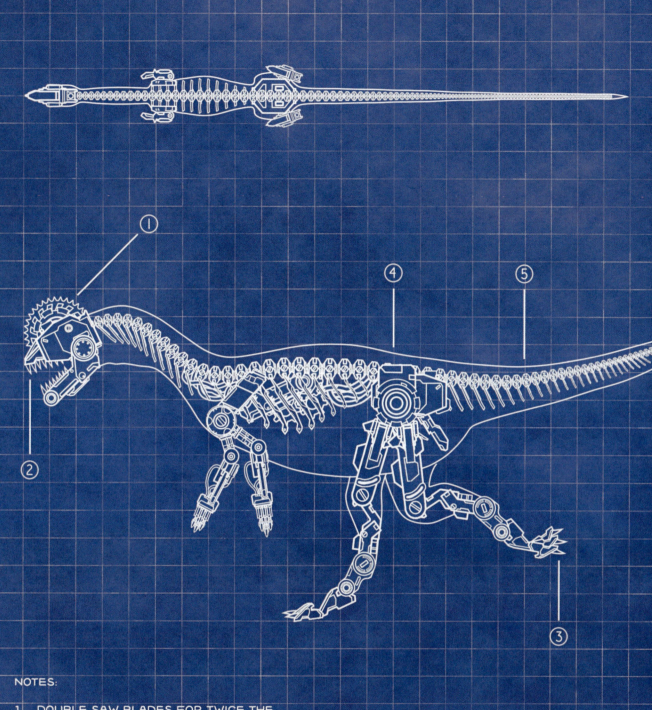

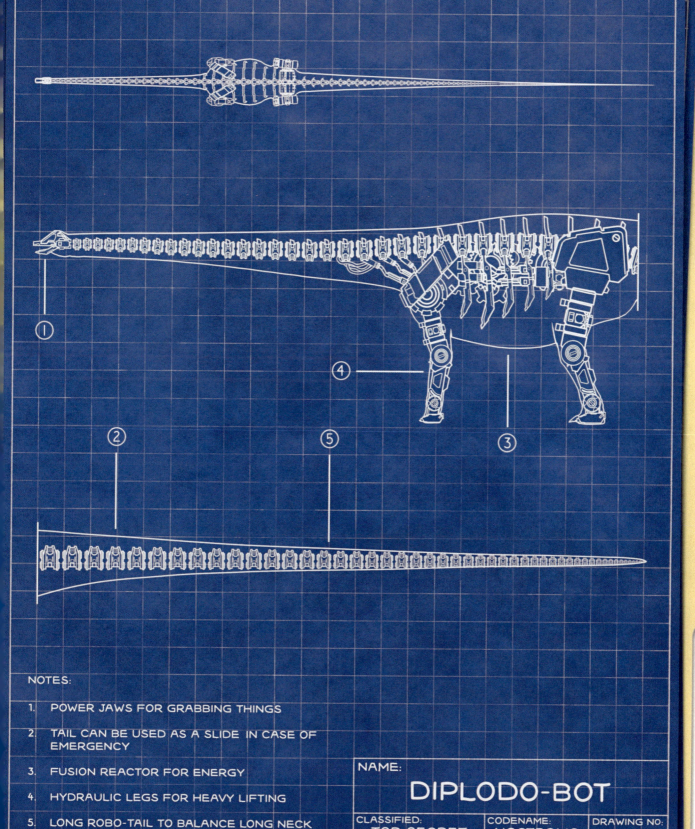

NOTES:

1. POWER JAWS FOR GRABBING THINGS
2. TAIL CAN BE USED AS A SLIDE IN CASE OF EMERGENCY
3. FUSION REACTOR FOR ENERGY
4. HYDRAULIC LEGS FOR HEAVY LIFTING
5. LONG ROBO-TAIL TO BALANCE LONG NECK

NAME:		
DIPLODO-BOT		
CLASSIFIED: TOP SECRET	CODENAME: NOSTROMO	DRAWING NO: 1 OF 15
LEAD ENGINEER: P. WEYLAND	CHK BY: PROF. INKENBLAUT	
DESCRIPTION: MIXED-USE ROBOTIC DIPLODOCUS		REV: D.8

DIPLODOCUS

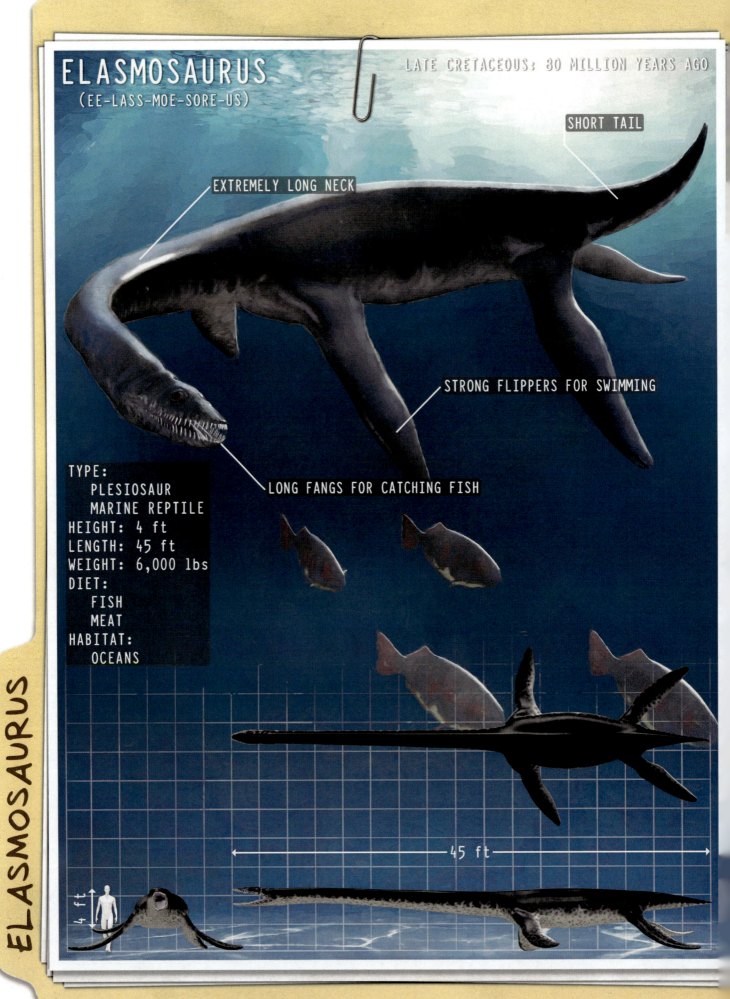

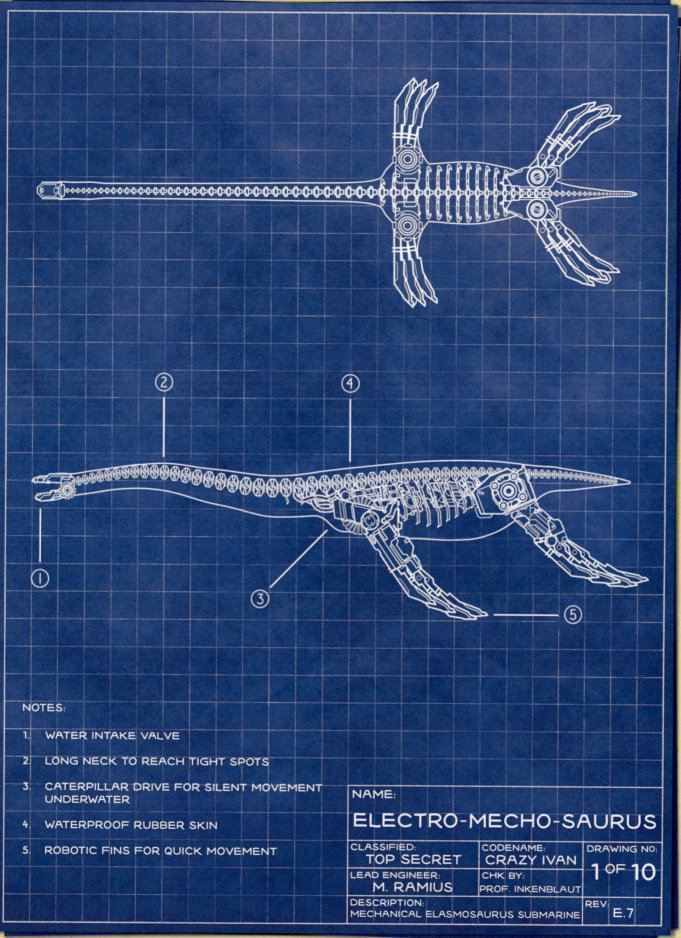

NOTES:

1. WATER INTAKE VALVE
2. LONG NECK TO REACH TIGHT SPOTS
3. CATERPILLAR DRIVE FOR SILENT MOVEMENT UNDERWATER
4. WATERPROOF RUBBER SKIN
5. ROBOTIC FINS FOR QUICK MOVEMENT

NAME:		
ELECTRO-MECHO-SAURUS		
CLASSIFIED: TOP SECRET	CODENAME: CRAZY IVAN	DRAWING NO: 1 OF 10
LEAD ENGINEER: M. RAMIUS	CHK BY: PROF. INKENBLAUT	
DESCRIPTION: MECHANICAL ELASMOSAURUS SUBMARINE		REV. E.7

ELASMOSAURUS

IGUANODON
(IG-WAN-OH-DON)

EARLY CRETACEOUS: 126 MILLION YEARS AGO

- SHARP BEAK FOR PICKING AT PLANTS
- STRONG BACK LEGS
- SPECIALIZED THUMB SPIKE
- COULD WALK ON TWO LEGS OR FOUR

TYPE: DINOSAUR
HEIGHT: 12 ft
LENGTH: 33 ft
WEIGHT: 6,000 lbs
DIET: PLANTS
HABITAT: WOODLANDS

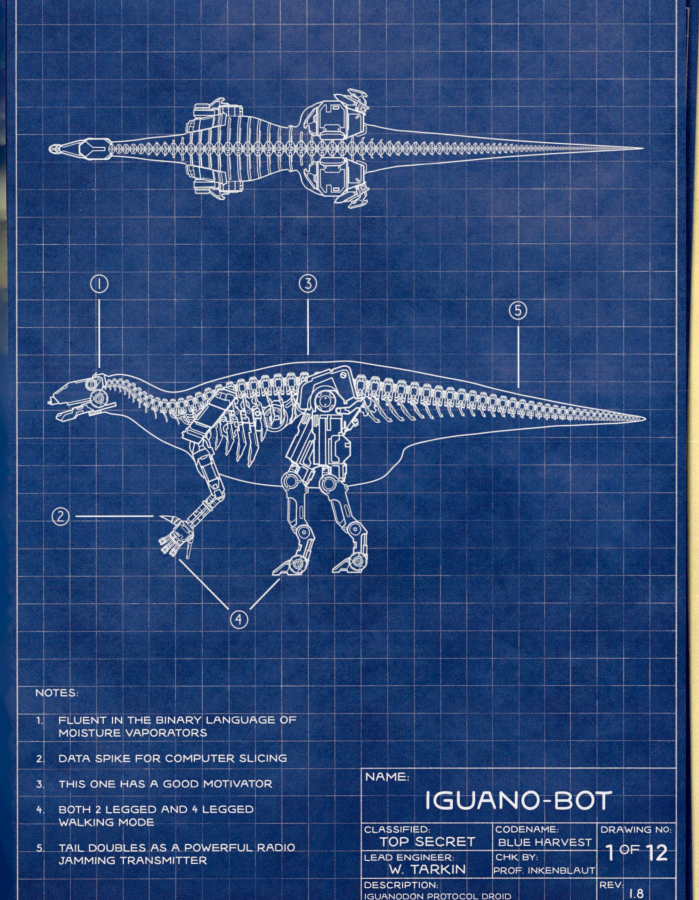

PACHYCEPHALOSAURUS
(PACK-EE-SEFF-AH-LOW-SORE-US)

LATE CRETACEOUS: 70 MILLION YEARS AGO

- 10 IN. THICK BONY DOME ON SKULL
- SPIKES AROUND DOMED SKULL
- SMALL BEAK
- HEAVY TAIL

TYPE: DINOSAUR
HEIGHT: 5 ft
LENGTH: 15 ft
WEIGHT: 1,000 lbs
DIET: PLANTS
HABITAT: FORESTS

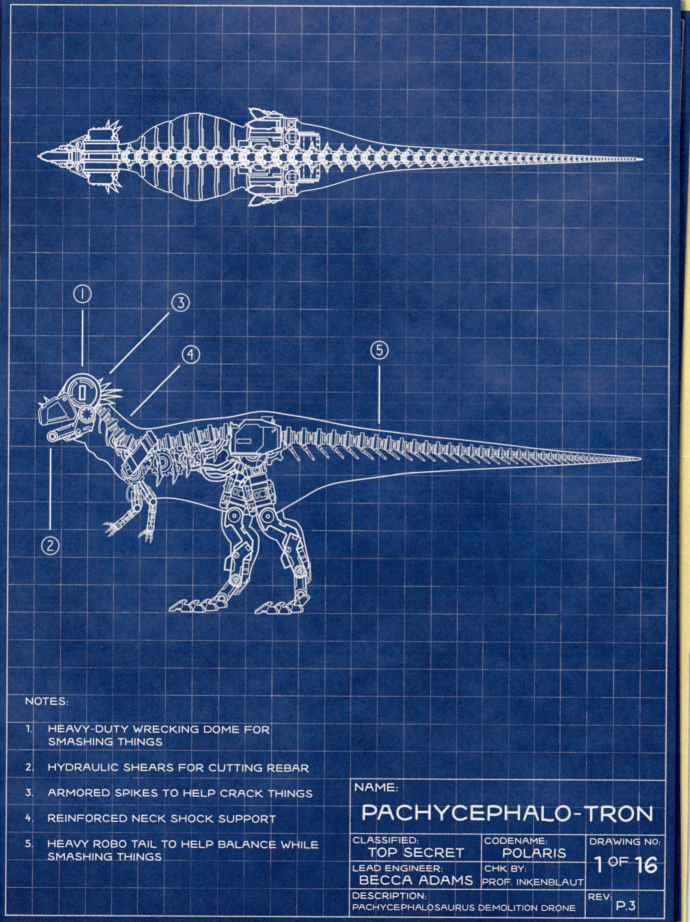

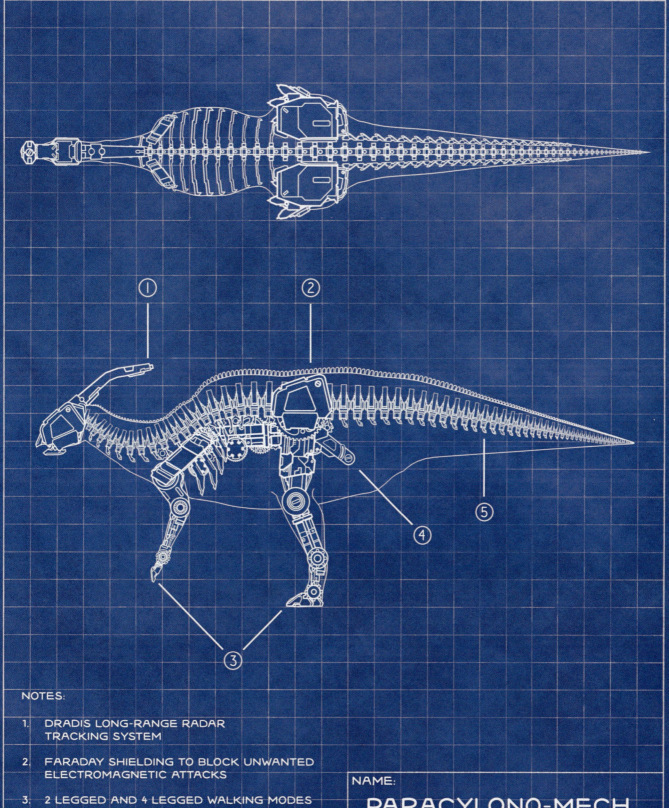

NOTES:

1. DRADIS LONG-RANGE RADAR TRACKING SYSTEM
2. FARADAY SHIELDING TO BLOCK UNWANTED ELECTROMAGNETIC ATTACKS
3. 2 LEGGED AND 4 LEGGED WALKING MODES
4. RADAR SIGNAL AMPLIFIER
5. SOUND DAMPENER FOR QUIET MOVEMENT

NAME: PARACYLONO-MECH		
CLASSIFIED: TOP SECRET	CODENAME: CAPRICA	DRAWING NO: 1 OF 10
LEAD ENGINEER: W. ADAMA	CHK BY: PROF. INKENBLAUT	
DESCRIPTION: LONG-RANGE RADAR PARASAUROLOPHUS ROBOT		REV: P.4

PARASAUROLOPHUS

PROTOCERATOPS
(PRO-TOE-SERRA-TOPS)

LATE CRETACEOUS: 75 MILLION YEARS AGO

- LARGE NECK FRILL
- VERY LARGE EYES
- SHORT LEGS
- BEAK WITH A POWERFUL BITE

TYPE: DINOSAUR
HEIGHT: 2 ft
LENGTH: 6 ft
WEIGHT: 400 lbs
DIET: PLANTS
HABITAT: DESERT

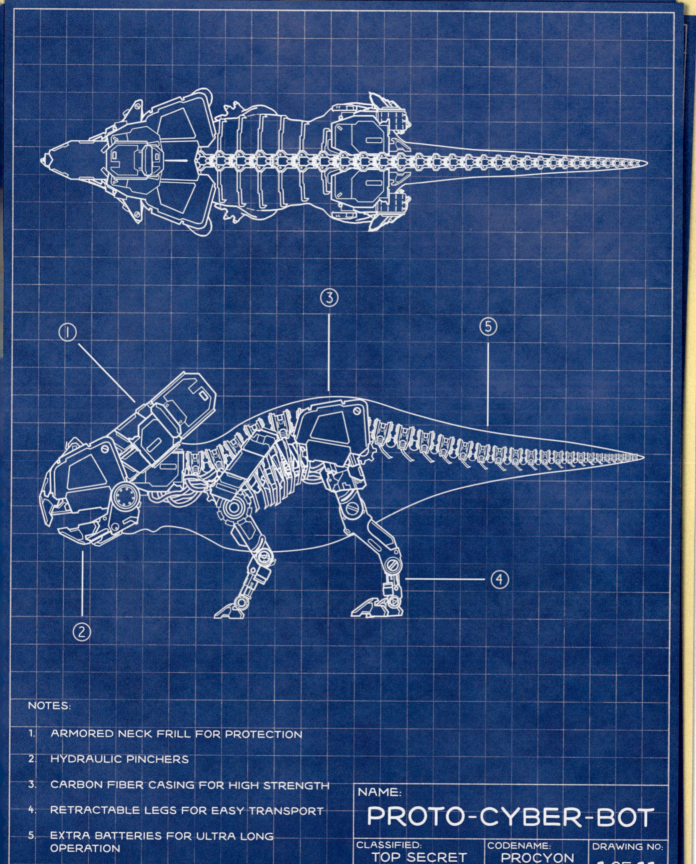

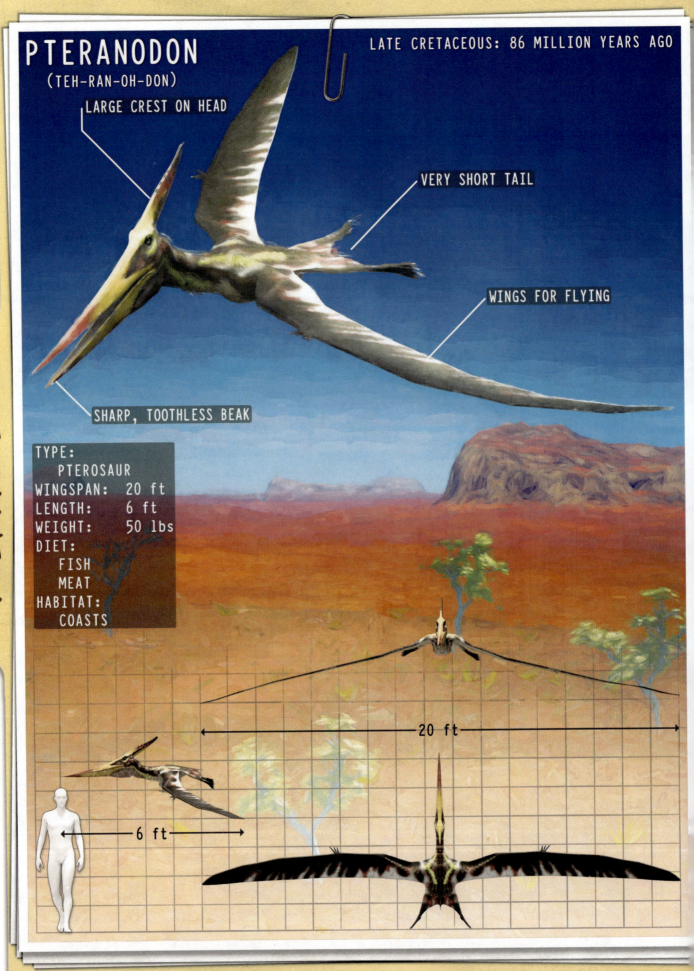

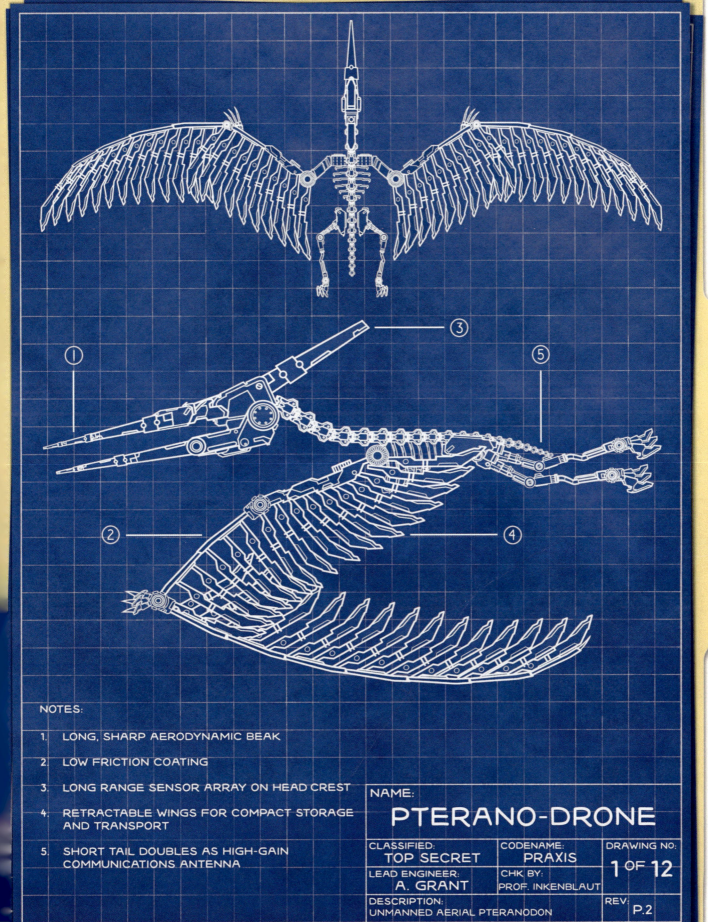

PTERODACTYLUS
(TERR-OH-DAK-TIL-US)

LATE JURASSIC: 150 MILLION YEARS AGO

- CREST ON HEAD
- VERY SHORT TAIL
- WINGS FOR FLYING
- SMALL SHARP TEETH

TYPE:
    PTEROSAUR
WINGSPAN: 3 ft
LENGTH: 2 ft
WEIGHT: 10 lbs
DIET:
    FISH
    MEAT
HABITAT:
    COASTAL AREAS

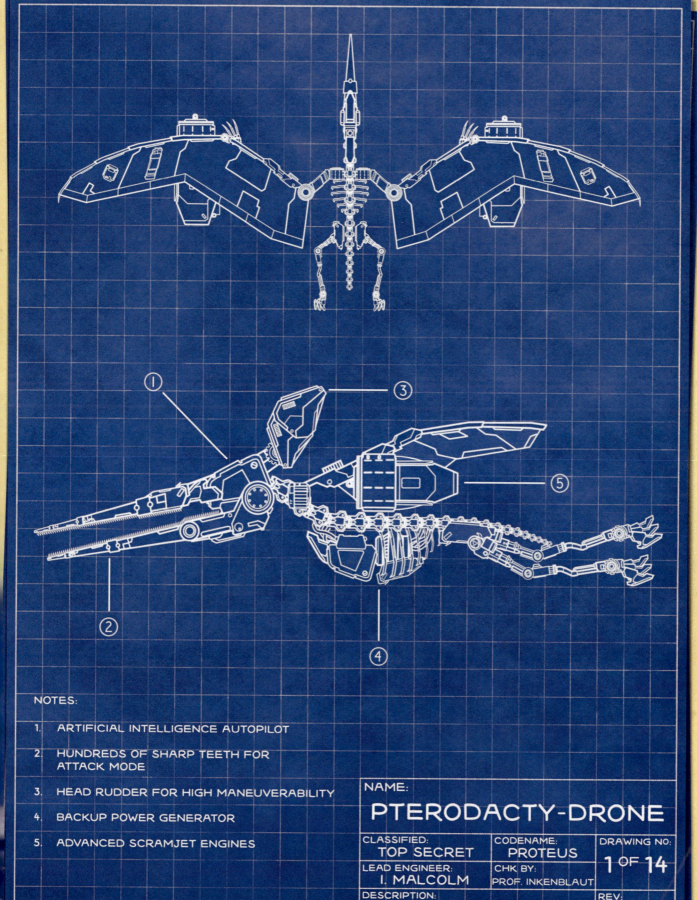

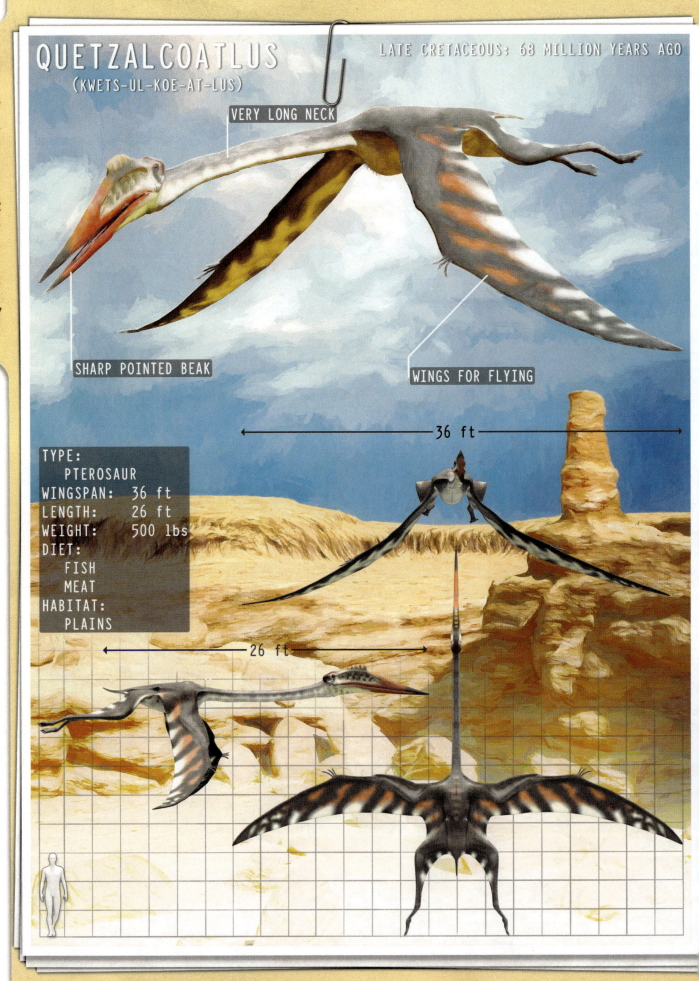

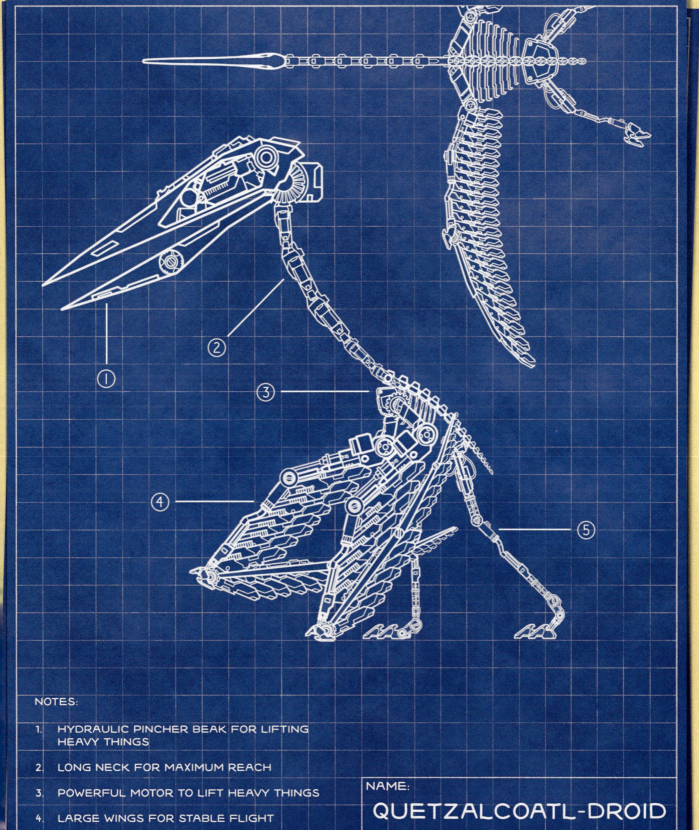

NOTES:

1. HYDRAULIC PINCHER BEAK FOR LIFTING HEAVY THINGS
2. LONG NECK FOR MAXIMUM REACH
3. POWERFUL MOTOR TO LIFT HEAVY THINGS
4. LARGE WINGS FOR STABLE FLIGHT
5. EXTENDABLE LEGS FOR GROUND WORK

NAME:		
QUETZALCOATL-DROID		
CLASSIFIED: TOP SECRET	CODENAME: MULTIPASS	DRAWING NO: **1 OF 13**
LEAD ENGINEER: LEELOO DALLAS	CHK BY: PROF. INKENBLAUT	
DESCRIPTION: QUETZALCOATLUS HEAVY AIRLIFT CRANE		REV: Q.8

QUETZALCOATLUS

SARCOSUCHUS
(SAR-KOH-SOO-KUSS)

EARLY CRETACEOUS: 112 MILLION YEARS AGO

- COVERED IN ARMORED BONE PLATES
- POWERFUL TAIL FOR SWIMMING
- POWERFUL JAWS
- MOUTH FULL OF SHARP TEETH
- SHORT LEGS

TYPE: CROCODILE-LIKE REPTILE
HEIGHT: 4 ft
LENGTH: 39 ft
WEIGHT: 16,000 lbs
DIET: FISH, MEAT
HABITAT: SWAMPS

39 ft

4 ft

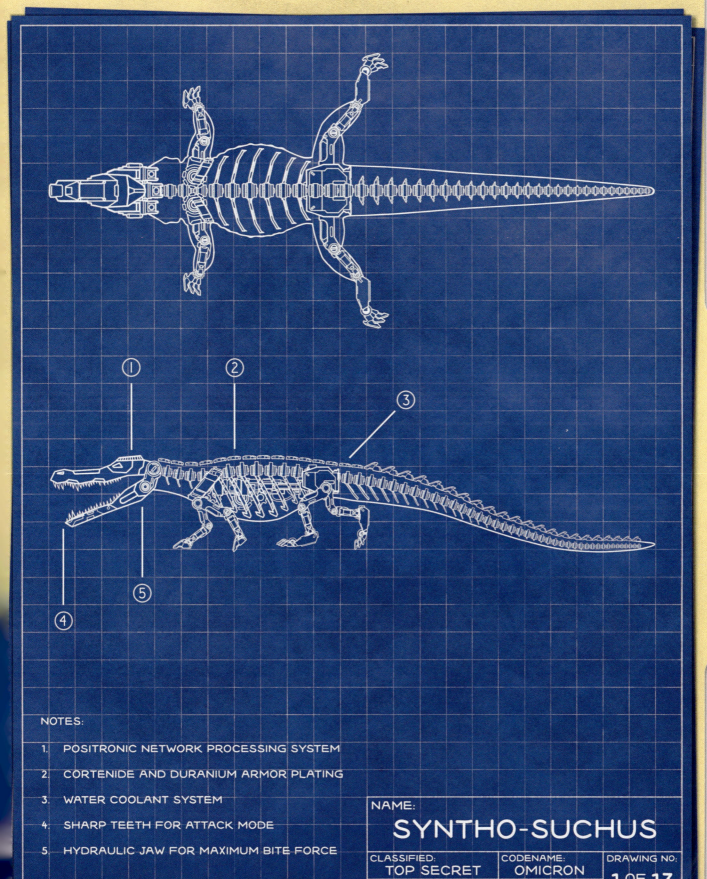

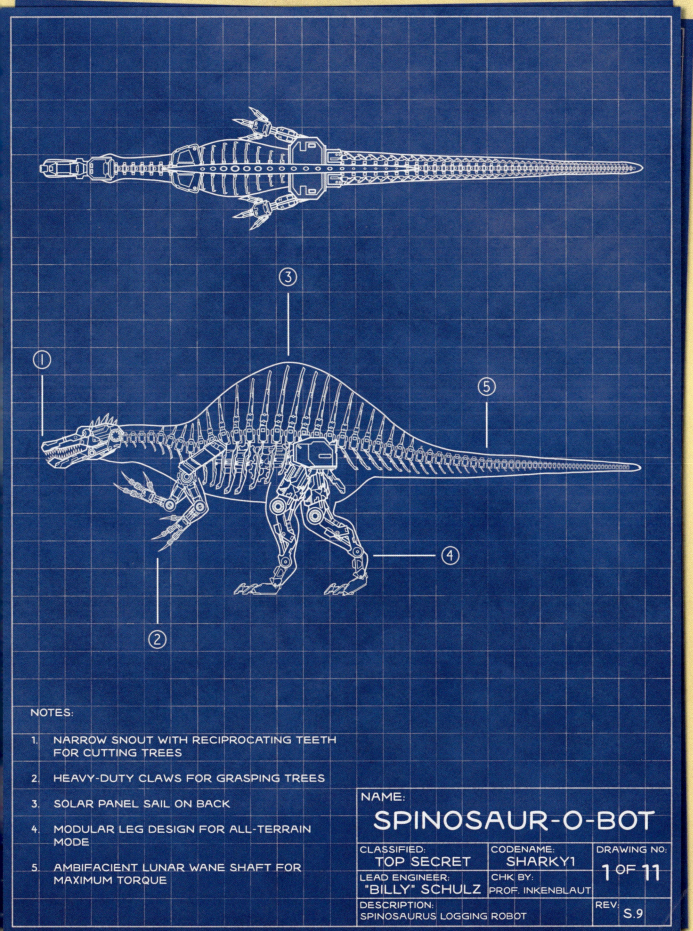

NOTES:
1. NARROW SNOUT WITH RECIPROCATING TEETH FOR CUTTING TREES
2. HEAVY-DUTY CLAWS FOR GRASPING TREES
3. SOLAR PANEL SAIL ON BACK
4. MODULAR LEG DESIGN FOR ALL-TERRAIN MODE
5. AMBIFACIENT LUNAR WANE SHAFT FOR MAXIMUM TORQUE

NAME: **SPINOSAUR-O-BOT**

CLASSIFIED: TOP SECRET
CODENAME: SHARKY1
DRAWING NO: 1 OF 11
LEAD ENGINEER: "BILLY" SCHULZ
CHK BY: PROF. INKENBLAUT
DESCRIPTION: SPINOSAURUS LOGGING ROBOT
REV: S.9

SPINOSAURUS

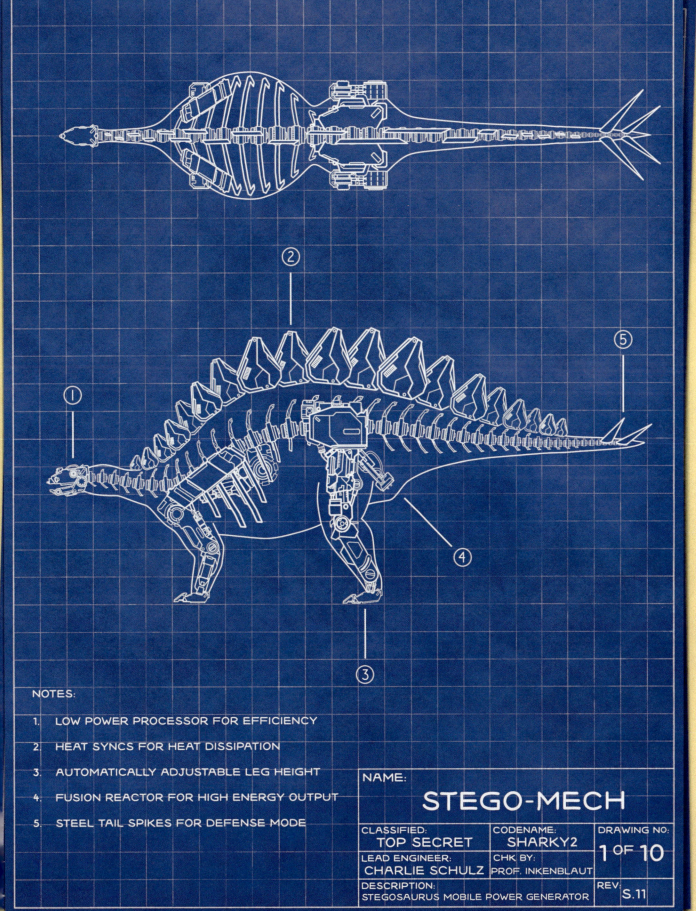

STYRACOSAURUS
(STY-RACK-OH-SORE-US)

LATE CRETACEOUS: 75 MILLION YEARS AGO

- NECK FRILL SURROUNDED BY LONG HORNS
- SINGLE HORN AT END OF SNOUT
- SHORT TAIL
- TOOTHLESS BEAK
- FOUR SHORT LEGS

TYPE:
 DINOSAUR
HEIGHT: 9 ft
LENGTH: 18 ft
WEIGHT: 6,000 lbs
DIET:
 PLANTS
HABITAT:
 WOODLANDS

18 ft

9 ft

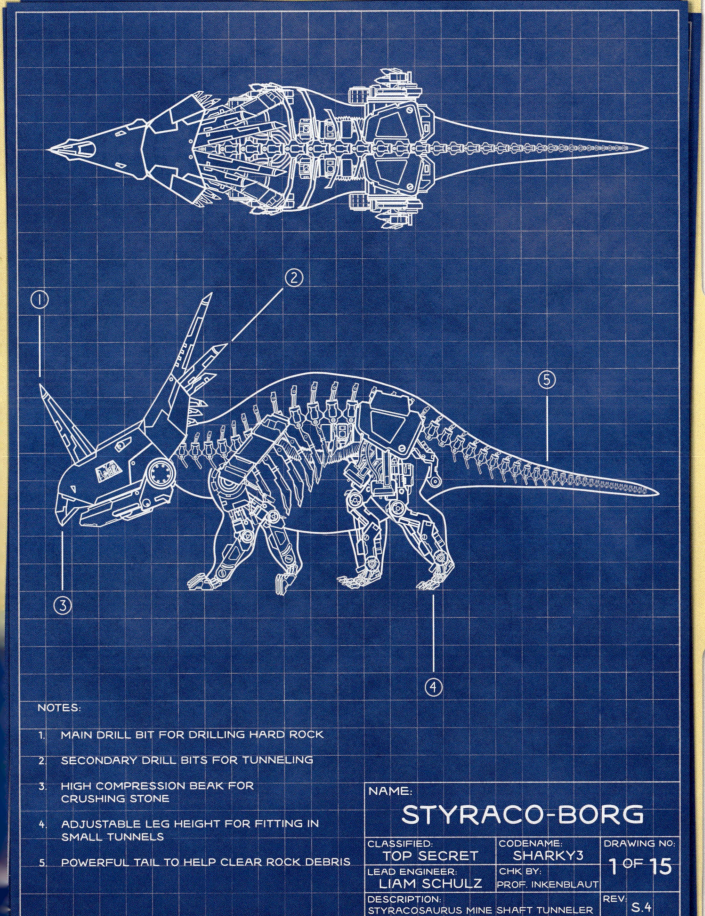

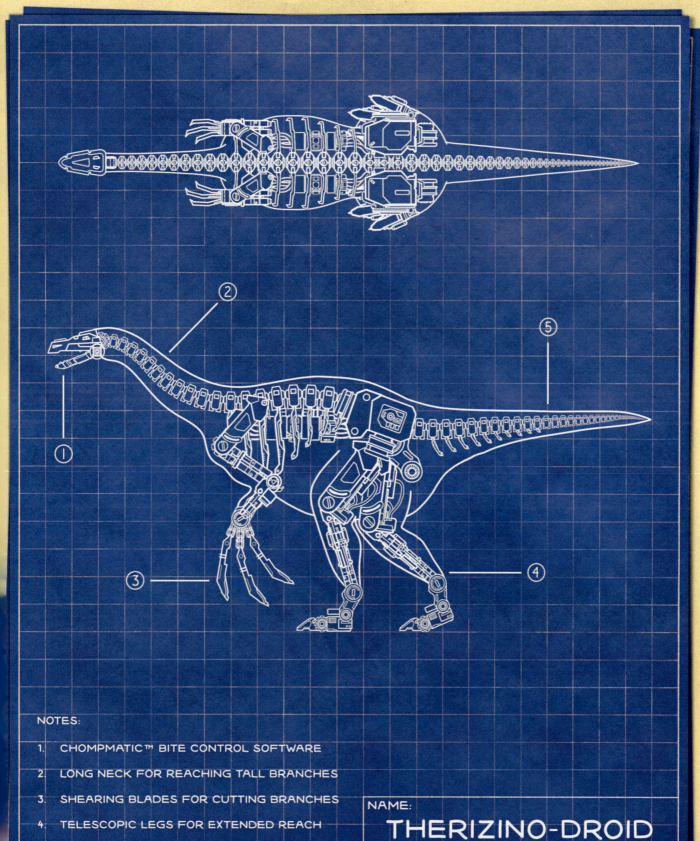

NOTES:

1. CHOMPMATIC™ BITE CONTROL SOFTWARE
2. LONG NECK FOR REACHING TALL BRANCHES
3. SHEARING BLADES FOR CUTTING BRANCHES
4. TELESCOPIC LEGS FOR EXTENDED REACH
5. ROBOTIC TAIL CAN BE USED FOR BALANCE WHEN REACHING

NAME:		
THERIZINO-DROID		
CLASSIFIED: TOP SECRET	CODENAME: FREDDY	DRAWING NO: 1 OF 13
LEAD ENGINEER: DR SHERRI FORD	CHK BY: PROF. INKENBLAUT	
DESCRIPTION: THERIZINOSAURUS TREE PRUNING BOT		REV: T.4

THERIZINOSAURUS

TRICERATOPS
(TRY-SERRA-TOPS)

LATE CRETACEOUS: 68 MILLION YEARS AGO

- LONG HORNS FOR DEFENSE
- SOLID NECK FRILL
- SHORT TAIL
- TOOTHLESS BEAK
- POWERFUL LEGS

TYPE: DINOSAUR
HEIGHT: 10 ft
LENGTH: 30 ft
WEIGHT: 24,000 lbs
DIET: PLANTS
HABITAT: WOODLANDS

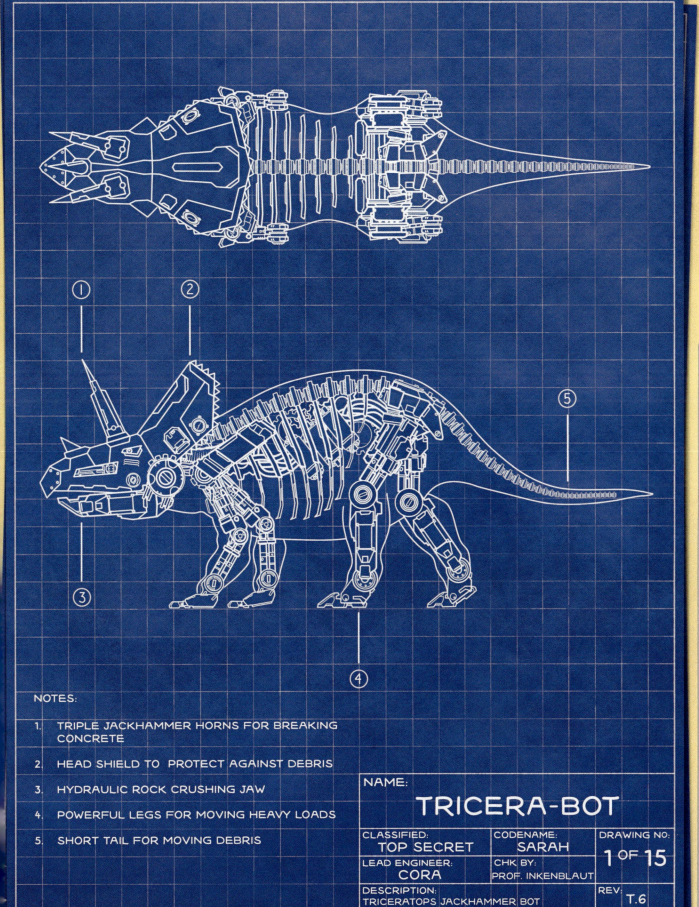

TRICERATOPS

NOTES:

1. TRIPLE JACKHAMMER HORNS FOR BREAKING CONCRETE
2. HEAD SHIELD TO PROTECT AGAINST DEBRIS
3. HYDRAULIC ROCK CRUSHING JAW
4. POWERFUL LEGS FOR MOVING HEAVY LOADS
5. SHORT TAIL FOR MOVING DEBRIS

NAME:		
TRICERA-BOT		
CLASSIFIED: TOP SECRET	CODENAME: SARAH	DRAWING NO:
LEAD ENGINEER: CORA	CHK BY: PROF. INKENBLAUT	1 OF 15
DESCRIPTION: TRICERATOPS JACKHAMMER BOT		REV. T.6

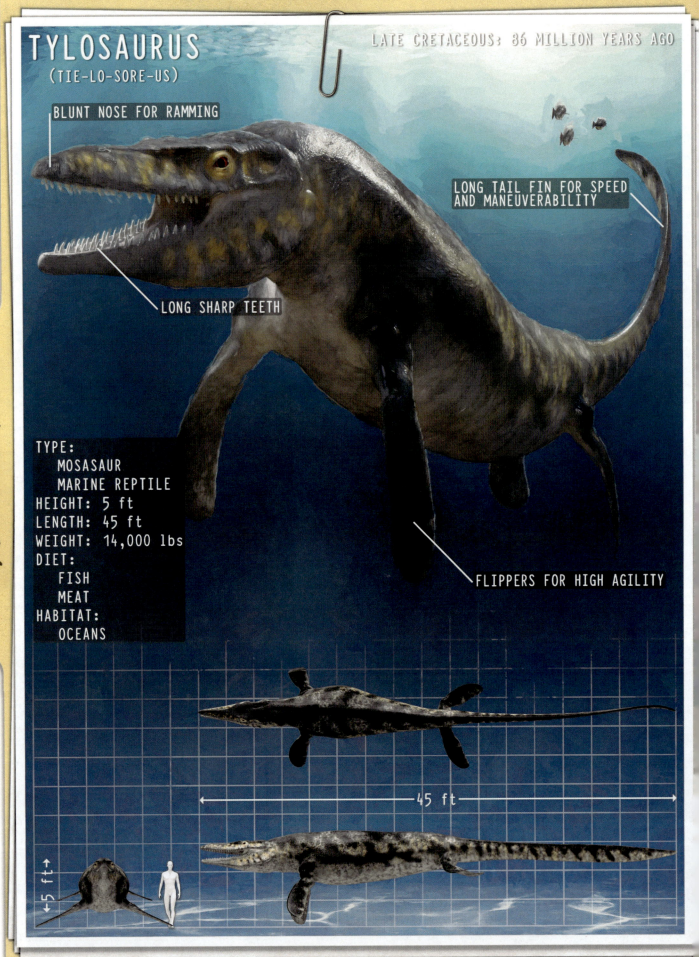

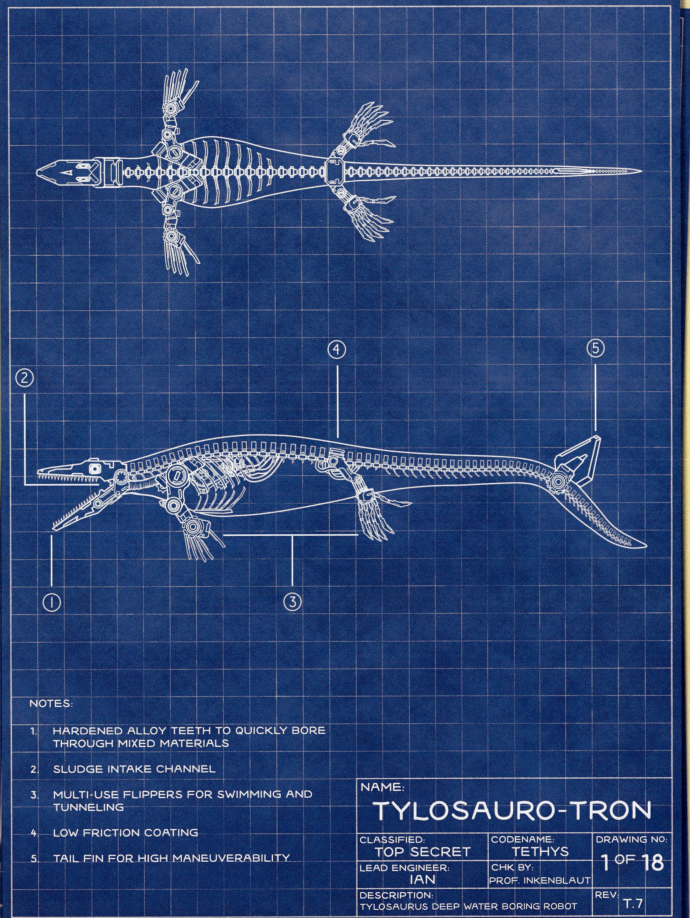

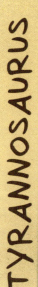

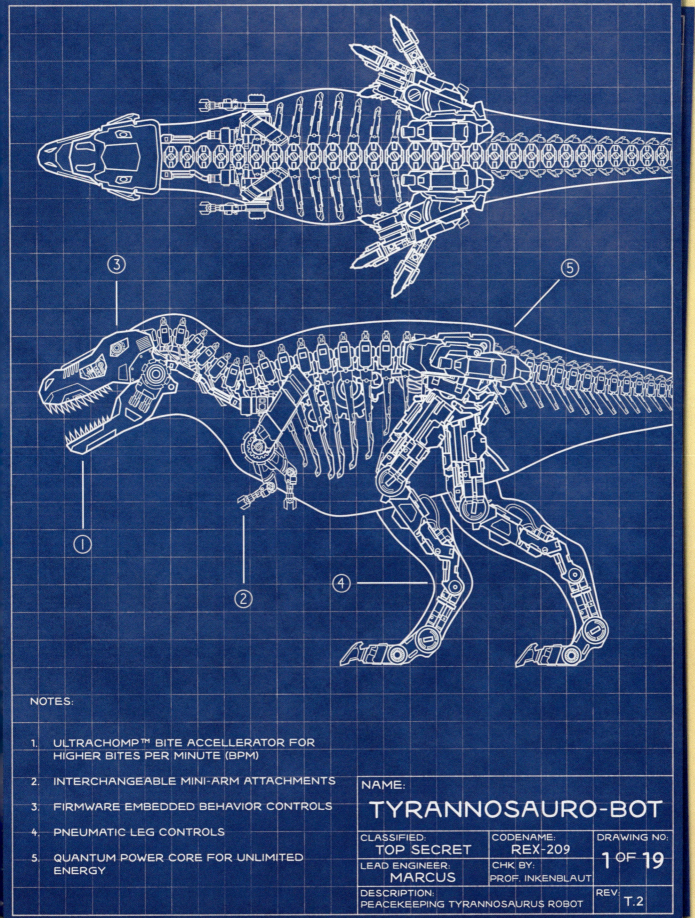

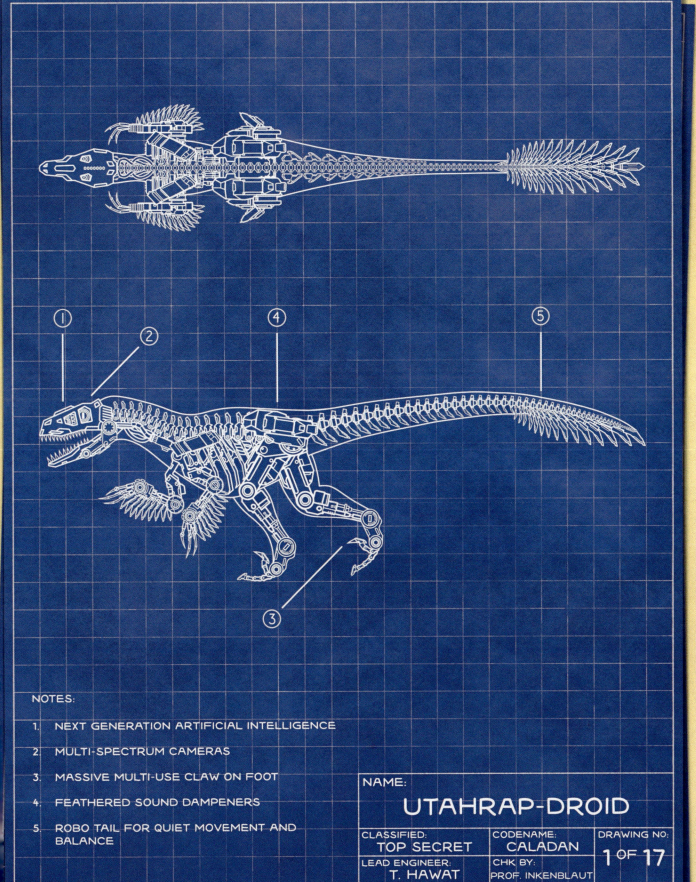

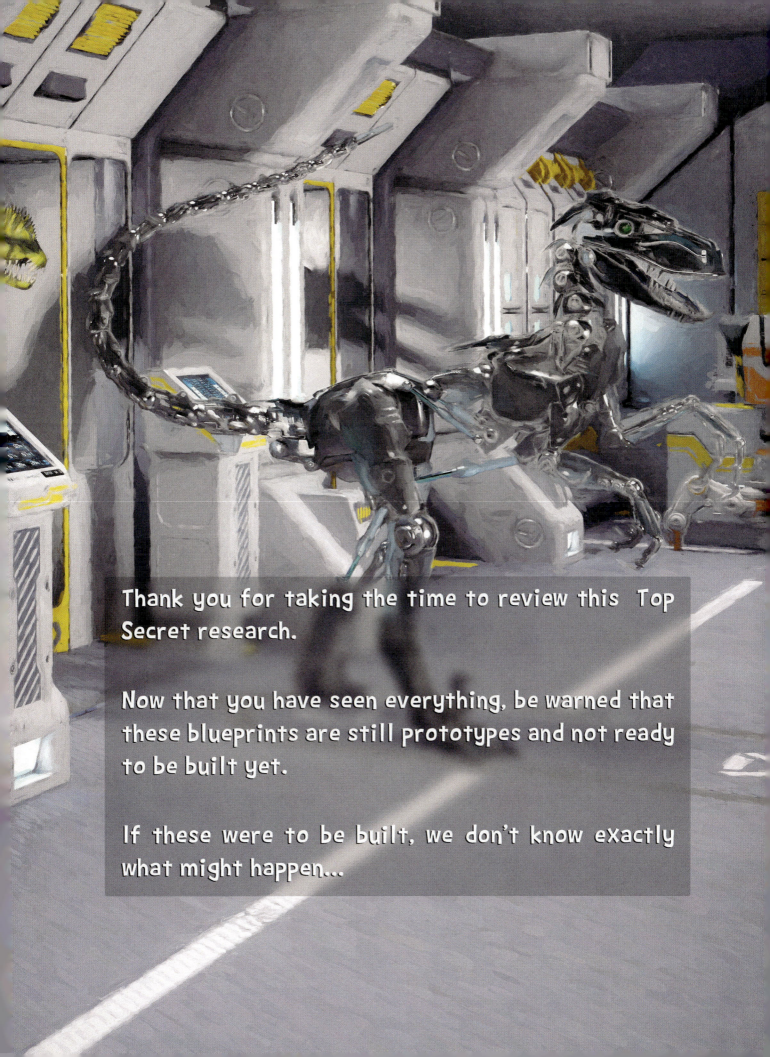

WWW.BEDTIME.PRESS

FOR MORE INFORMATION ON UPCOMING PROJECTS, GIVEAWAYS AND OTHER CHANCES TO GET FREE BOOKS, FOLLOW @BEDTIMEPRESS ON SOCIAL MEDIA.

COPYRIGHT © 2018 BY DAVID CUNLIFFE

ALL RIGHTS RESERVED. THIS BOOK OR ANY PORTION THEREOF MAY NOT BE REPRODUCED OR USED IN ANY MANNER WHATSOEVER WITHOUT THE EXPRESS WRITTEN PERMISSION OF THE AUTHOR.

PUBLISHED BY BEDTIME PRESS
WWW.BEDTIME.PRESS